NATIONAL
GEOGRAPHIC
KiDS

美 国 国 家 地 理
双 语 阅 读

Saving Animal Babies
救助动物宝宝

懿海文化 编著

马鸣 译

第三级

外语教学与研究出版社
FOREIGN LANGUAGE TEACHING AND RESEARCH PRESS
北京 BEIJING

京权图字：01-2021-5130

Original title: Saving Animal Babies
Copyright © 2013 National Geographic Partners, LLC. All Rights Reserved.
Copyright © 2022 Bilingual Simplified Chinese/English edition National Geographic Partners, LLC.
All Rights Reserved.
NATIONAL GEOGRAPHIC and Yellow Border Design are trademarks of the National Geographic Society, used under license.

图书在版编目 (CIP) 数据

救助动物宝宝：英文、汉文／懿海文化编著；马鸣译. —— 北京：外语教学与研究出版社，2021.11（2023.8 重印）
（美国国家地理双语阅读. 第三级）
书名原文：Saving Animal Babies
ISBN 978-7-5213-3147-9

Ⅰ. ①救… Ⅱ. ①懿… ②马… Ⅲ. ①动物－少儿读物－英、汉 Ⅳ. ①Q95-49

中国版本图书馆 CIP 数据核字 (2021) 第 228170 号

出 版 人 王　芳
策划编辑 许海峰 刘秀玲 姚　璐
责任编辑 姚　璐
责任校对 华　蕾
装帧设计 许　岚
出版发行 外语教学与研究出版社
社　　址 北京市西三环北路 19 号（100089）
网　　址 https://www.fltrp.com
印　　刷 天津海顺印业包装有限公司
开　　本 650×980 1/16
印　　张 37.5
版　　次 2022 年 3 月第 1 版 2023 年 8 月第 4 次印刷
书　　号 ISBN 978-7-5213-3147-9
定　　价 188.00 元（全 15 册）

如有图书采购需求，图书内容或印刷装订等问题，侵权、盗版书籍等线索，请拨打以下电话或关注官方服务号：
客服电话：400 898 7008
官方服务号：微信搜索并关注公众号"外研社官方服务号"
外研社购书网址：https://fltrp.tmall.com

物料号：331470001

Table of Contents

The Cubs Are Coming!

It is a dark and quiet night at the zoo. The tiger is restless. The zookeeper thinks the tiger will have her babies soon. And she does.

Tiger babies are called cubs.

There are four new tigers in the world! Tigers are in danger of becoming extinct. That means every tiger is special.

Wild Word
EXTINCT: A type of plant or animal no longer living

Eat, Sleep, Repeat!

This is the tiger's first litter of cubs. Some tigers don't take care of their first litter. Without help, the cubs could die. But caretakers at the zoo know what to do.

At first, the cubs only need to eat and sleep. Every three hours, the cubs drink warm milk.

The cubs have a blanket that smells like their mother. Sleep well, little tigers.

Wild Word
LITTER: A group of animals born at one time

Would you like to try a chunky meat milk shake? The cubs are crazy for them.

First the veterinarian checks the cubs' baby teeth. They need to be strong and sharp to chew the chunks. Then the zoo chef buys jars of turkey baby food. He mixes it with milk and vitamins to make the milk shake.

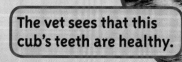

The vet sees that this cub's teeth are healthy.

A tiger cub is weighed to make sure it's growing well.

Now the cubs will grow even faster. One day they will be full-grown tigers.

Wild Word
VETERINARIAN: A doctor for animals, called vet for short

A Long, Tall Baby

Molly is three days old. She is an 80-pound, 5-foot-tall baby giraffe. This baby should drink three gallons of milk a day. But her mother cannot make milk for her.

Molly also has an infection. She needs help. She has to go to the hospital.

Wild Word
INFECTION: A sickness caused by a virus or bacteria

The vet feeds Molly milk. The tube in Molly's neck is held in place by a bandage.

The vet puts goat's milk in a giant baby bottle. She has to hold the bottle up high for Molly.

The vet puts a little tube in Molly's neck. It doesn't hurt and is an easy way to give Molly the medicine she needs.

Soon Molly is better and back with her mom.

Molly is healthy and growing. She is almost as tall as her mother!

Wanted: One Hairy Mom

Remy's mother got sick before he was born. She could not take care of Remy.

Orangutans need to be raised by other orangutans. Remy needed a foster mother to take care of him.

Madu is a grown-up orangutan. She never had a baby of her own.

But she had cared for two other orangutan babies that didn't have moms. Would Madu be a foster mother to Remy, too?

Wild Word

FOSTER MOTHER: An animal that is not family but cares for a young animal like a mom

Remy is an orangutan baby. Young orangutans stay with their moms for five to seven years.

Remy snuggles with Madu.

With his blanket and toys, Remy went to meet Madu. It was love at first sight. Soon Remy climbed on Madu's back.

Remy watched Madu. Madu taught Remy what to eat. She showed him how to hang and climb. Remy learned how to be an orangutan.

Q How do you catch an ape?

A Climb a tree and act like a banana!

Madu teaches Remy how to hang with one arm.

17

Toys for Tots

All babies love toys. Zookeepers try to get the right toy for each baby.

Sloth babies cling to their moms. A stuffed pillow works, too.

Even the youngest monkeys can learn to hang from ropes or chains.

Elephants love splashing and swimming in water. A kiddie pool is lots of fun!

Polar bear cubs like to chase and pounce on a ball.

A treat frozen in ice is a puzzle for curious panda cubs.

Young tigers like to play with each other. Another tiger is better than any toy.

Saving a
Seal Pup

Guinness is a gray seal, just like this one.

Even ocean babies need help sometimes.

Wildlife rescuers saw a seal pup on the beach. Seals leave the water to rest. But this little guy was too thin. He didn't go back in the water. He was in trouble.

Rescuers wrapped him in a wet towel and took him to the hospital. They named him Guinness.

Wild Word

WILDLIFE RESCUER: Someone who saves wild animals from danger

Guinness had a broken jaw. The vets operated and put a wire in his jaw. The wire held the jaw together while the bone healed.

Three months later, Guinness could eat by himself again!

Guinness enjoys a frozen fish in ice.

It was time to go back to the water.
Everyone cheered when Guinness
scooted back to the ocean where
he belonged.

Backyard Babies

Do you want to be a vet when you grow up?

You might want to be like Dr. Greg Mertz. People know him as the Odd Pet Vet. He takes care of all kinds of animals that need help.

Dr. Mertz helps a snake.

This three-month-old goose has a broken wing. The bandage works like a cast on a broken arm.

This painted turtle has a cracked shell. The bandage keeps away infection.

People bring hurt animals to Dr. Mertz. Many wild animals get hurt on roads. Luckily, Dr. Mertz can help most of the animals he sees.

Springtime is busy for Dr. Mertz.
That's when many babies are born.
Animal babies like to explore.
Sometimes they get into trouble
and need help.

Dr. Mertz to the rescue!

These opossum babies were
found in the wall of a house.

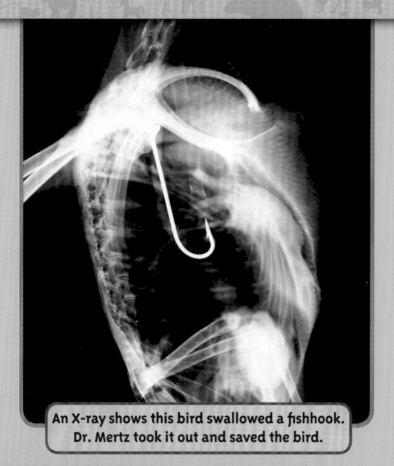

An X-ray shows this bird swallowed a fishhook.
Dr. Mertz took it out and saved the bird.

Raccoon cubs can live on their
own after three months. Until then,
Dr. Mertz keeps this one safe.

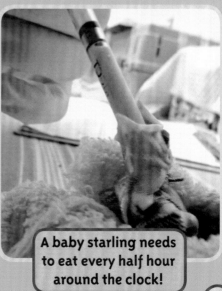

A baby starling needs
to eat every half hour
around the clock!

Dos and Don'ts

What can you do to help baby animals? Here are some dos and don'ts. . .

DO slide a ramp in a pool if you see baby frogs in it. Then they can climb out safely.

DON'T feed ducks and other birds bread. It's bad for them.

DO wait 24 hours before rescuing a baby deer or bunny. The mother is probably nearby.

DO tell an adult to call animal rescue if you see anyone hurting an animal.

DON'T adopt a wild animal. They do not make good pets.

DO prevent pets from harming wildlife. Put a bell on your cat's collar. Keep your dog on a leash.

DO pick up trash you see in the woods. Plastic bags and bottles can hurt animals.

DON'T pick up a baby bird that is on the ground. Ask an adult or call a vet or the local Audubon Society for advice.

Stump Your Parents

Can your parents answer these questions about baby animals? You might know more than they do!

Answers are at the bottom of page 31.

When tigers are born, they _____.

A. Drink milk
B. Are hungry for meat
C. Don't sleep
D. Sing

What do young orangutans learn from older orangutans?

A. How to find food
B. How to hang and climb
C. How to be an orangutan
D. All of the above

What should you do if you find a baby bird on the ground?

A. Run away
B. Leave it alone and tell an adult
C. Bring it home
D. Give it some candy

Wait 24 hours before rescuing a baby deer or bunny because _____.

A. It likes to be alone
B. It might be out getting a snack
C. The mother is probably nearby
D. It could be on its way to a party

Where does a seal pup live?

A. In the mountains
B. In the ocean
C. In the forest
D. In a department store

What do elephants love?

A. Chocolate
B. Reading
C. Water
D. Doing jumping jacks

What kind of milk can you feed a baby giraffe?

A. Chocolate milk
B. Goat's milk
C. Soy milk
D. Milk shakes

Answers: 1) A, 2) D, 3) B, 4) C, 5) B, 6) C, 7) B

Glossary

EXTINCT: A type of plant or animal no longer living

FOSTER MOTHER: An animal that is not family but cares for a young animal like a mom

INFECTION: A sickness caused by a virus or bacteria

LITTER: A group of animals born at one time

VETERINARIAN: A doctor for animals, called vet for short

WILDLIFE RESCUER: Someone who saves wild animals from danger

▶ 第4—5页

虎崽要出生啦！

这是动物园里一个幽暗、宁静的夜晚。老虎焦躁不安。动物园饲养员觉得老虎快要生宝宝了。她确实生了。

世界上多了四只老虎！老虎面临着灭绝的危险。这意味着每只老虎都很重要。

老虎宝宝叫"虎崽"。

野生动物小词典
灭绝：一种植物或动物不再存活

▶ 第6—7页

进食，睡觉，重复！

这是这只老虎的第一窝虎崽。有些老虎不照顾第一窝虎崽。如果没人帮忙，虎崽就会死掉。但动物园的饲养员知道该怎么做。

最初，虎崽只需要进食和睡觉。每隔三个小时，虎崽就要喝一些温热的奶。

虎崽有一块毯子，闻起来很像它们的妈妈。好好睡吧，小老虎。

野生动物小词典
一窝：同时出生的一群动物

▶ 第8—9页

你要不要来一杯大肉块奶昔？虎崽特别喜欢。

首先，兽医检查虎崽的乳牙。它们要结实、锋利才能嚼大块的肉。然后动物园的厨师买了很多罐火鸡肉做成的婴幼儿食品。他把这种食品和奶、维生素混合在一起，做成奶昔。

兽医看出这只虎崽的牙齿很健康。

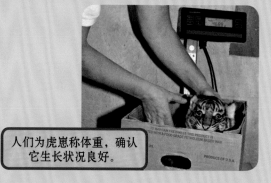

人们为虎崽称体重，确认它生长状况良好。

野生动物小词典

兽医：给动物看病的医生，英文简称vet

现在虎崽将会长得更快一些。总有一天它们会成为成年老虎。

▶ 第 10—11 页

又长又高的宝宝

莫莉 3 天大了。她是一个 80 磅（约 36.29 千克）重、5 英尺（约 1.52 米）高的长颈鹿宝宝。这个宝宝一天要喝 3 加仑（约 11.36 升）奶。但她的妈妈没有奶给她喝。

莫莉还被感染了。她需要帮助。她必须去医院。

野生动物小词典

感染：由病毒或细菌引发的疾病

▶ 第 12—13 页

兽医把山羊奶装到一个巨大的奶瓶里。她必须把奶瓶举得高高的，方便莫莉喝奶。

兽医把一个小药管放在莫莉的脖子里。它没有危害，而且这种方式很简便，可以让莫莉得到她需要的药。

莫莉很快就康复了，回到了妈妈的身边。

兽医喂莫莉喝奶。莫莉脖子里的药管用绷带固定好。

莫莉很健康，一天天地长大。她都快和妈妈一样高了！

▶ 第 14—15 页

招聘一位长毛妈妈

里米的妈妈在他出生前就生病了。她不能照顾里米。

猩猩需要别的猩猩抚养长大。里米需要一位养母来照顾他。

马杜是一只成年猩猩。她一直没有自己的孩子。

但她曾经照顾过另外两个没有妈妈的猩猩宝宝。马杜也会成为里米的养母吗？

> **野生动物小词典**
>
> 养母：不是一家人，却像妈妈一样照顾小动物的动物

里米是一个猩猩宝宝。年幼的猩猩要和妈妈一起生活5到7年。

▶ 第 16—17 页

里米带着他的毯子和玩具去见马杜。他们刚一见面就喜欢上了彼此。很快，里米便爬到了马杜的背上。

里米观察着马杜。马杜教里米应该吃什么。她向他展示如何吊挂、攀爬。里米学会了如何做一只猩猩。

里米依偎在马杜身旁。

马杜教里米如何单臂吊挂。

▶ 第 18—19 页

宝宝们的玩具

所有的宝宝都喜欢玩具。动物园饲养员试着为每个宝宝找到合适的玩具。

树懒宝宝总是紧紧地贴着它们的妈妈。填充枕头也可以。

北极熊幼崽喜欢追球和拍球玩。

即使是最年幼的猴子也能学会如何吊挂在绳子或链子上。

对好奇的大熊猫幼崽来说，冻在冰块里的美食是个智力玩具。

大象喜欢在水里戏水、游泳。儿童游泳池里很好玩儿！

年幼的老虎喜欢一起玩耍。老虎同伴胜过任何玩具。

▶ 第 20—21 页

救助小海豹

有时候，海生动物宝宝也需要帮助。

野生动物救援人员在海滩上发现了一只小海豹。海豹离开水上岸休息。但这个小家伙非常瘦。他没有回到水里。他遇到了麻烦。

救援人员用湿毛巾裹住他，把他送到了医院。他们给他取名叫"吉尼斯"。

吉尼斯是一只灰海豹，就像这只一样。

野生动物小词典

野生动物救援人员：把野生动物从危险中解救出来的人

▶ 第 22—23 页

吉尼斯在享用一条冷冻鱼。

吉尼斯的颌骨折了。兽医给他做手术，在他的颌里缝了一根金属线。在骨头愈合的过程中，这根线将他的颌固定在一起。

三个月后，吉尼斯又可以自己吃东西了！

是时候回到水里了。当吉尼斯快速地爬回属于他的海洋时，大家都欢呼起来。

▶ 第 24—25 页

后院里的宝宝

你长大后想当兽医吗？

你可能想成为像格雷格·默茨医生那样的人。人们都知道他是个古怪的宠物医生。他照顾各种各样需要帮助的动物。

人们把受伤的动物带到默茨医生那里。许多野生动物是在马路上受伤的。幸运的是，默茨医生能帮助他看到的大部分动物。

默茨医生帮助一条蛇。

这只三个月大的鹅翅膀断了一只。绷带就像固定在骨折的手臂上的石膏一样发挥作用。

这只锦龟的壳裂了。绷带可以防止感染。

▶ 第 26—27 页

对默茨医生来说，春天是个忙碌的季节。许多动物宝宝在这个时候出生。动物宝宝喜欢探索。有时候他们会遇到麻烦，需要帮助。

默茨医生来救援！

这些负鼠宝宝是在一座房子的墙里被发现的。

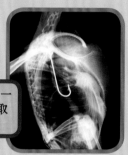

X光片显示这只鸟吞下了一个鱼钩。默茨医生把它取出来，救了这只鸟。

浣熊幼崽出生三个月后才能独立生活。在此之前，默茨医生保护着它的安全。

椋鸟宝宝每隔半小时就要吃一次东西，昼夜不停！

▶ 第 28—29 页

行为准则

你可以做些什么来帮助动物宝宝呢？这里有一些行为准则……

如果你在人造坡道上看到青蛙宝宝，就悄悄地把人造坡道放到池塘里。然后它们可以安全地爬出来。

不要给鸭子和其他鸟类喂面包吃。这对它们不好。

在救助鹿宝宝或兔宝宝之前要等待24小时。它们的妈妈可能就在附近。

如果看到有人伤害动物，让大人给动物救援机构打电话。

不要收养野生动物。它们不适合做宠物。

阻止宠物伤害野生动物。在你的小猫的项圈上系个铃铛。为你的狗拴上皮带。

拾起你在树林里看到的垃圾。塑料袋和塑料瓶会伤害动物。

不要捡地上的鸟宝宝。问大人该怎么办，也可以给兽医或当地的奥杜邦协会打电话，听听他们的建议。

挑战爸爸妈妈

　　你的爸爸妈妈能回答这些有关动物宝宝的问题吗？你知道的可能比他们还多呢！答案在第 31 页下方。

1 老虎出生时，它们 ＿＿＿＿＿＿。
A. 喝奶　　B. 很想吃肉　　C. 不睡觉　　D. 唱歌

2 年幼的猩猩能从成年猩猩身上学到什么？
A. 怎样找到食物　　B. 怎样吊挂和攀爬
C. 怎样做一只猩猩　　D. 以上都是

3 如果你在地上发现一个鸟宝宝，你应该怎么做？
A. 跑开　　　　B. 把它留在那里，告诉大人
C. 把它带回家　　D. 给它点儿糖果

4 在救助鹿宝宝或兔宝宝之前要等待 24 小时，因为
＿＿＿＿＿＿。
A. 它喜欢自己待着　　　B. 它可能是出来觅食的
C. 它的妈妈可能在附近　　D. 它可能在去派对的路上

5 小海豹在哪里生活？
A. 在大山里　　B. 在海洋里　　C. 在森林里　　D. 在百货商店里

6 大象喜爱什么？
A. 巧克力　　B. 阅读　　C. 水　　D. 跳跃运动

7 你可以喂长颈鹿宝宝什么奶？
A. 巧克力奶　　B. 山羊奶　　C. 豆奶　　D. 奶昔

词汇表

灭绝：一种植物或动物不再存活

养母：不是一家人，却像妈妈一样照顾小动物的动物

感染：由病毒或细菌引发的疾病

一窝：同时出生的一群动物

兽医：给动物看病的医生，英文简称vet

野生动物救援人员：把野生动物从危险中解救出来的人